Balance and Motion

Developed at
The Lawrence Hall of Science,
University of California, Berkeley
Published and distributed by
Delta Education,
a member of the School Specialty Family

© 2012 by The Regents of the University of California. All rights reserved. No part of this book may be reproduced or transmitted in any form or by any means, electronic or mechanical, including photocopying or recording, or by any information storage and retrieval system, without permission in writing from the publisher.

1325241
978-1-60902-034-7
Printing 4 — 12/2012
Quad/Graphics, Leominster, MA

Table of Contents

Make It Balance! 3
Push or Pull? 13
Things That Spin 18
Rolling, Rolling, Rolling! 23
Strings in Motion 30
Move It, but Don't Touch It 36
Tools and Machines 41
Glossary 47

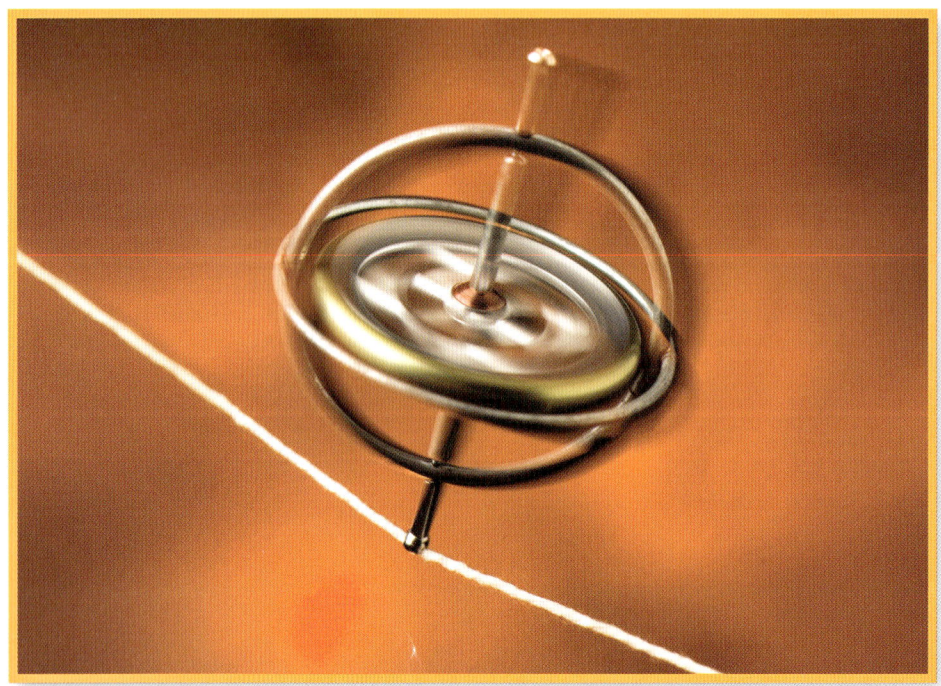

Make It Balance!

We live in a world full of **motion**.
But not everything moves in the same way.

Some things move from one place to another.
Some things **spin** around and around.

Other things **balance**.
They might move only if you give them a little **push**.

At the circus, you can see people balancing in amazing ways.
They might balance on ropes or on chairs.
It takes lots of practice.

With some practice, you can balance on a log.
Is it easier with your arms down by your sides or with your arms out?

Try balancing on one foot.
Is it easier with your eyes open or closed?

In some countries, people balance things on their heads.
This is how they carry things from place to place.

7

What can you balance on your head?
Try a book or an apple. How far can you walk
and keep it balanced?

What will happen to this toy if you gently push on it?

It might wobble.

But it will return to where it started.

It will return to a **stable position**.

This toy is **counterbalanced** to help it come back to balance, even if it is pushed.

You see things balancing all around you. Which pictures show things that are balancing?

What keeps things in a stable position?

Thinking about Make It Balance!

1. Think about balancing on one foot. What can you do with your body to help you balance?

2. Is it easier to balance a ball or a book on your head? Why?

3. What does *balance* mean?

4. What does *counterbalance* mean?

Push or Pull?

One way to make things move is to push them.
Another way to make things move is to **pull** them.

A push or a pull is called a **force**.
You always need a force to make something happen.
A push or a pull can make something move, stop it from moving, or change its direction.

You can push things with your hands or body. Moving air can also push things.

When you play baseball, do you use the bat to push or pull the ball?

It doesn't matter if you hit a home run or strike out. **Gravity** always pulls the ball to the ground. Gravity is a pulling force.

Think about a roller coaster.
What pulls it up the first hill?
What force pulls it back down again?

Push or pull, a force is always needed to make things move.

16

Thinking about Push or Pull?

1. What are two things a force can do to an object's motion?

2. Tell about one way you can move a ball. Is the force a push or a pull?

3. Tell how to spin a pinwheel, and describe the force.

4. Think about throwing a ball into the air. What will gravity do to this ball?

Things That Spin

Things that spin are all around us.
When something spins, it turns on its **axis**.

Tops need to spin fast to balance.
What happens when a top slows down?

Some things spin slowly, like this Ferris wheel.

Look at these pictures.

Which things spin fast?
Which things spin slowly?
Which things don't spin at all?

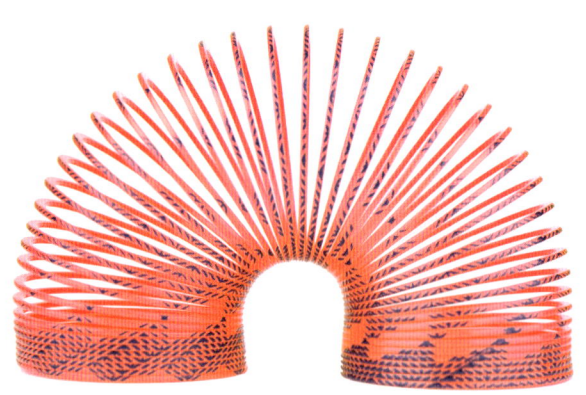

21

Thinking about Things That Spin

1. What does *spin* mean?

2. Name four things that spin.

3. Name three things that don't spin.

4. How does a top stay balanced?

Rolling, Rolling, Rolling!

Things can move by rolling.
When something **rolls**, it goes around and around.
But instead of staying in one spot, it moves from one place to another.

Things that have round surfaces roll easily.
Marbles roll, and so do cans.

Look at the wheels on this page.
Are the wheels spinning or rolling?
What's the difference between spinning and rolling?

Some things don't roll in a straight line.
Things that are shaped like a paper cup don't roll straight.

Some things don't roll at all.
Things that are flat won't roll.

Which of these things will roll down the ramp?

Gravity helps things roll downhill.
It's fun to see how fast something will go.

Which ramp would you use?
Why?

Ready, set, go!

Things move in different ways.
Spinning things go around and around.
Rolling things go around and ahead.

This toy bear can roll and balance.
Can you see why?

Fast, slow, up, down, spinning, rolling, sliding . . .
Almost everything moves!

What ways can you move?

Thinking about Rolling, Rolling, Rolling!

1. Name four things that roll.

2. Describe what happens to a tennis ball when it rolls on the floor.

3. Why can't a block roll down a ramp?

4. What is the difference between rolling and spinning?

Strings in Motion

Some musical instruments have strings. Guitars, harps, and violins are stringed instruments.
The strings move, and you hear **sounds**.
How do moving strings make sounds?

A harp's strings move back and forth when you pluck them.
Back-and-forth motion is called **vibration**.
Vibrating strings push on the air.
This motion makes sound waves.
The waves move out in all directions.
When sound waves enter your ears, you hear sound.

Short strings vibrate faster than long strings.
Strings that vibrate fast make high sounds.
They sound like ting, ping, bing.

Long strings vibrate more slowly than short strings.
Strings that vibrate slowly make low sounds.
They sound like boom, thum, dom.

A harp can make very soft sounds, like a whisper. Or it can make big, loud sounds, like a yell. Soft and loud are **properties** of sound called **volume**.

A harp player changes the volume of the sound by how hard she plucks each string.
She chooses the **pitch** by plucking strings of different lengths.
When the strings move in just the right way, you hear beautiful music.

Which one of these stringed instruments makes high-pitched sounds?
Which one makes low-pitched sounds?

Thinking about Strings in Motion

1. What happens when you pluck strings on a harp?

2. What do you call the back-and-forth motion of strings?

3. Would you pluck a short string or a long string to make a low-pitched sound?

4. Do long strings vibrate fast or slowly?

Move It, but Don't Touch It

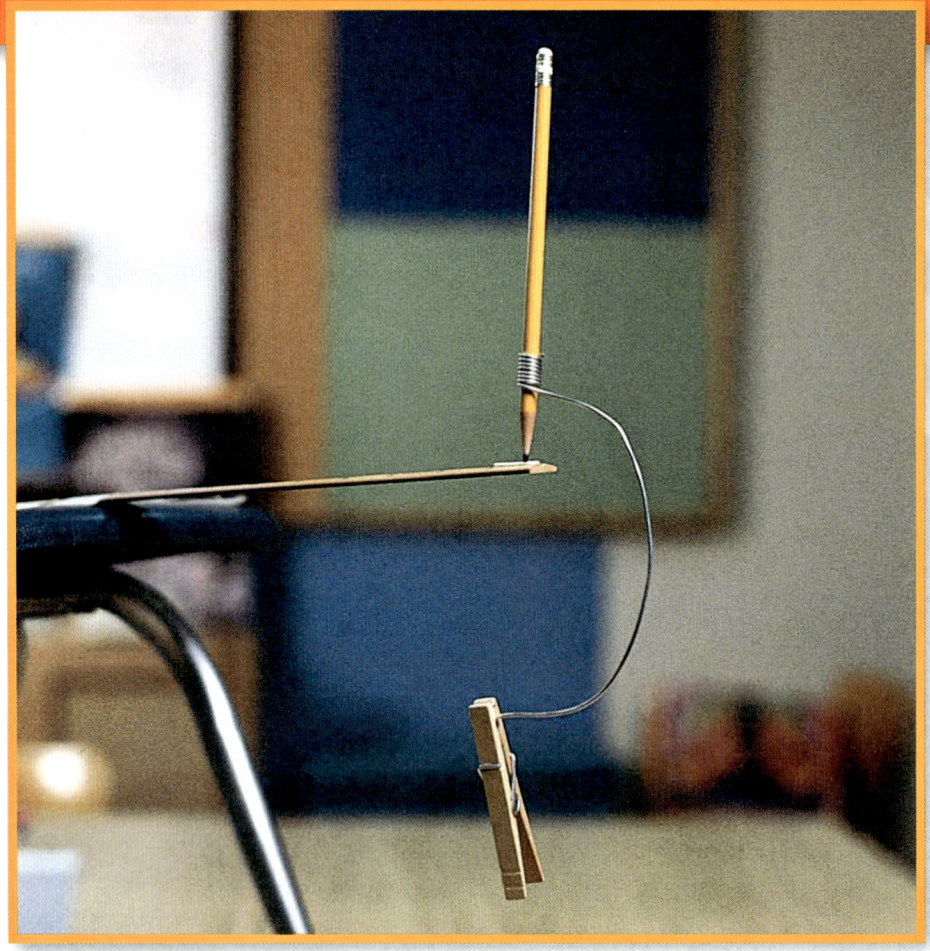

A pencil balanced on its point is in a stable position.
You can move it with a little push.
But can you move it without touching it?
Try a **magnet**.

Magnetism is a force.
The magnetic force can push and pull other magnets.
One magnet can make another magnet move.
The magnets don't even have to touch!

Magnets also pull on metals like iron and steel.
Look at the pencil balanced on its point.
The clothespin is used as a **counterweight**.
The spring in the clothespin is steel.
A magnet can make the balanced pencil move back and forth.
The magnet never touches the pencil.

Here's another way to use a magnet to move something without touching it.
Hang a paper clip from a string.
Use a magnet to lift the paper clip.
How high can you lift the paper clip without touching it?

A **compass** is a magnetic needle in a case.
The colored end of the needle points north.
Bring another magnet close to the compass.
The needle will turn.
The needle moves without being touched.

The magnetic force works through air to make things move.

39

Thinking about Move It, but Don't Touch It

1. Think about a paper clip on top of a table. Tell about three ways to move the paper clip.

2. What is a way to move the paper clip without touching it?

3. Name some things that magnets move.

4. Is magnetism a force? Why or why not?

Tools and Machines

People use **tools** and **machines** to help them do work.

You can't push a nail into a board with your hands. But you can use a tool to help.

A hammer is a tool that makes it easy to push nails into wood.
A hammer uses force to push nails into wood.

Hammers can pull with force, too.
You can use a hammer to remove a nail from wood.

42

Nuts and bolts hold things together.
Nuts and bolts keep wheels on cars.
Bolts with nuts hold bridges together.
Nuts and bolts must stay together and hold tight.

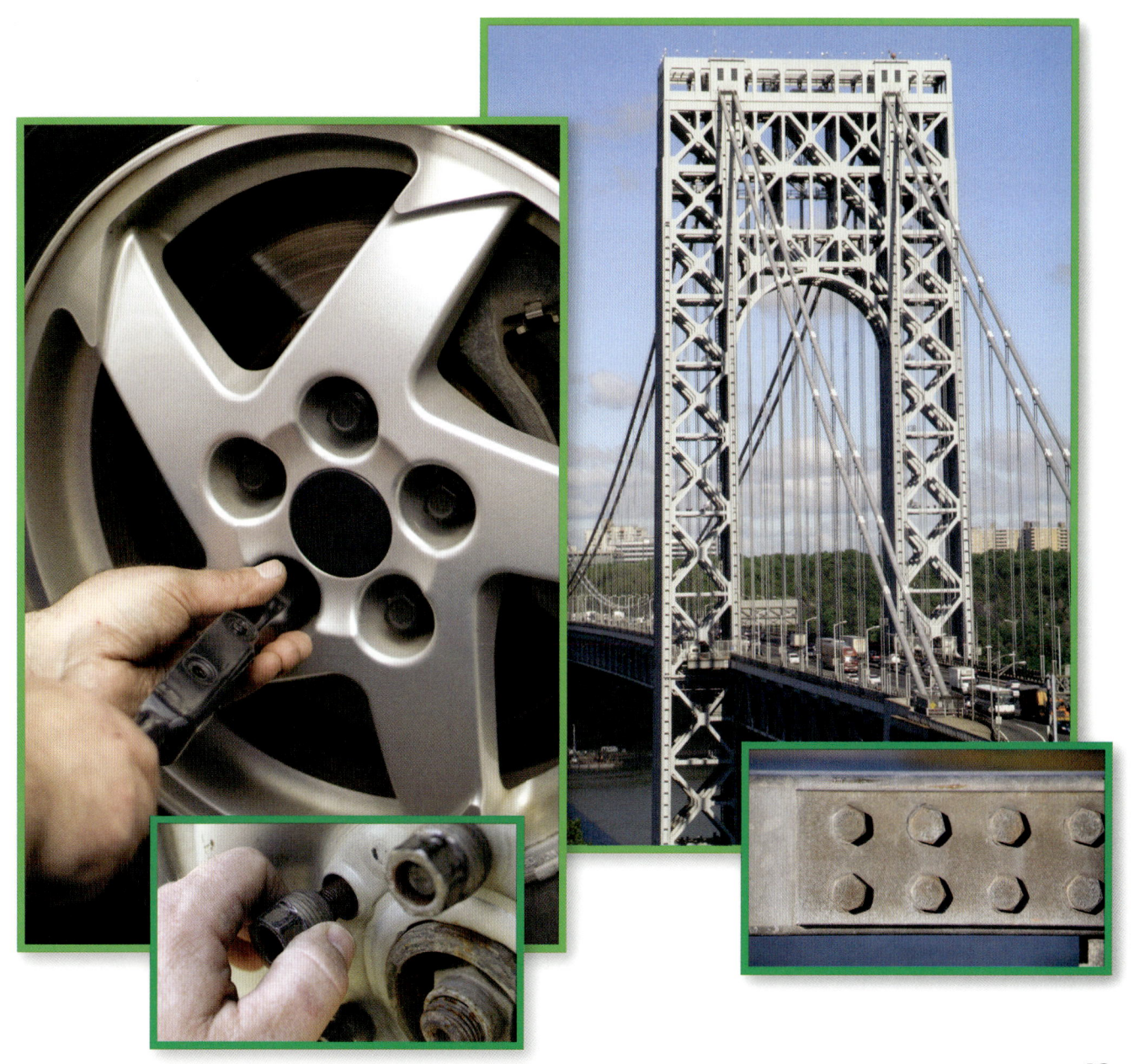

43

A wrench is a tool.
You can use it to turn nuts, bolts, and screws.
A wrench tightens nuts, bolts, and screws that are too hard to turn by hand.

With a wrench, you can use more force to tighten nuts, bolts, and screws.

How do people move really heavy things?

They use machines.

Forklifts and trucks are machines.

This forklift uses a lot of force to lift a heavy load.

The forklift makes it easy to move the load from place to place.

People use machines to move things when a lot of force is needed.

Thinking about Tools and Machines

1. Name some tools that help people do work.

2. What kind of work can you do with a hammer and nail?

3. What type of force will remove a nail from wood?

4. Name some machines that help people do work.

Glossary

axis a straight line around which something turns **(18)**

balance to be in a stable position **(3)**

compass a magnetic needle in a case. Compass needles on Earth point north. **(39)**

counterbalance to place weights on an object to keep it in a stable position **(9)**

counterweight weight added to a system to make it balance **(37)**

force a push or a pull **(13)**

gravity a force that pulls things toward Earth **(15)**

machine something that helps people do work. A machine can make heavy things move by using a push or a pull. **(41)**

magnet an object that sticks to iron and steel. Magnets can push or pull other magnets or objects made of iron or steel. **(36)**

magnetism a force that can work on another object without touching it **(36)**

motion the act of moving **(3)**

pitch how high or low a sound is **(34)**

property something you can observe about an object or a material. Volume is an example of a property of sound. **(33)**

pull when you make things move toward you. Pulling is a force. **(13)**

47

push when you make things move away from you. Pushing is a force. **(3)**

roll to move from one place to another by turning over and over **(23)**

sound something you hear **(30)**

spin to move by turning around an axis **(3)**

stable position steady, not falling over **(9)**

tool something that helps people do work. A tool can make things move by using a push or a pull. **(41)**

vibration a back-and-forth motion. Vibration makes sound. **(31)**

volume how soft or loud a sound is **(33)**